ASYMMETRIC ELECTROSTATIC FORCES AND A NEW ELECTROSTATIC GENERATOR

COMPUTER SCIENCE, TECHNOLOGY AND APPLICATIONS

Additional books in this series can be found on Nova's website at:

https://www.novapublishers.com/catalog/index.php?cPath=23_29&
seriesp=Computer+Science%2C+Technology+and+Applications

Additional E-books in this series can be found on Nova's website at:

https://www.novapublishers.com/catalog/index.php?cPath=23_29&
seriesp=Computer+Science%2C+Technology+and+Applications

COMPUTER SCIENCE, TECHNOLOGY AND APPLICATIONS

ASYMMETRIC ELECTROSTATIC FORCES AND A NEW ELECTROSTATIC GENERATOR

KATSUO SAKAI

Novinka
Nova Science Publishers, Inc.
New York

NOTICE TO THE READER

LIBRARY OF CONGRESS CATALOGING-IN-PUBLICATION DATA

Available upon Request
ISBN: 978-1-61728-920-0

Published by Nova Science Publishers, Inc. ✛ *New York*

CONTENTS

Preface   vii

Chapter 1   Asymmetric Force (AF)   1

Chapter 2   The Asymmetric Image Force (AIF)   19

Chapter 3   The Asymmetric Coulomb Force   25

Chapter 4   A New Electrostatic Generator
             (An Application of the ACF)   41

Chapter 5   A Miracle Charge Carrier that Can
             Move Forward in a Reverse Field   55

Chapter 6   Conclusion   63

Appendix 1   An Explanation of the Simulation Method
             (A Bi-Dimensional Axi-Symmetric
             Finite Difference Method)   65

Appendix 2   Relationship between the Approximate Formula
             for the Gradient Force and the Simulation
             Method of the Electrostatic Force Acting on a
             Non-Charged Cylinder in a Convergent Field   69

References   75

Index   77

PREFACE

When a charge q is placed in a uniform electric field E, electrostatic force F that acts on this charge can be calculated by Coulomb's law (F=qE). This is a well known formula. Almost all of electrostatic problems are solved by this formula. Many times we forget the limitation condition of this formula. Almost all of text books write that Coulomb's law (This formula) is only correct to point charge. Some text books add that Coulomb's law (This formula) is correct to charged hollow conductive sphere too.

Today, electrostatics deals with many charged powders. A charged powder can be replaced with a point charge in large space. Therefore, we have solved almost all electrostatic problems by Coulomb's law (This formula) for a long time.

I have been researching new methods of electrostatic generator on this ten years. In first five years, I used sphere shape for charge carrier. Unfortunately, I could not develop effective method. In recent five years, I used asymmetric shape, for example cup, for charge carrier. I succeeded in developing effective methods. In this development, I simulated electrostatic force that acts on charged charge carrier having cup shape. The simulated results were different from the results that were calculated with this formula. I wanted to confirm my simulation results with official textbook or paper. But, I can not find them. Only the following text book answers my question a little.

The original title of this book is "Fundamentals of PHYSICS". Three authors of this book were Halliday, Resnick and Walker. This book was published in 2001 by John Wiey & Sons.Inc.. I heard that this is a very famous reference book for the freshman in the U.S.A. The subject of chapter 22 of this book (Japanese version) is electric charge. Coulomb's law is explained in this chapter. There are 11 problems at the end of this chapter.

The first one is "Is Coulomb's law correct with all charge carriers?" The answer is "No, it is only correct with charged particle and hollow sphere."

This answer does not confirm the results of my simulation, but it does not refuse the results of my simulation. I think how to solve electrostatic force that acts on asymmetric shape charge carrier is unknown field. This must be the suburbs of coulomb's law.

I think Coulomb's law is similar to Stokes law. We can easily solve the air resistance for small sphere with Stokes formula. But, we must execute a lot of wind tunnel experiment and/or simulations before we chose best shape of a new airplane. In the same way, I must execute a lot of electrostatic experiment and/or simulations before I chose best shape of a charge carrier of the new electrostatic generator. .

I have done many simulations of asymmetric electrostatic forces for this purpose. And, I have done some experiments of asymmetric electrostatic forces. I will tell you main results of them in this book. And, I will tell you about the new electrostatic generator. This generator can solve the environmental problem and the energy crisis at the same time.

I think you will get a question about my opinion after you read this book. Because of, the theory of the new generator looks to be against the law of the conservation of energy. But, it is not against the law of the conservation of energy. I think electrostatic potential energy of many charge in asymmetric shape charge carrier is different to electrostatic potential energy of single charge (point charge). But, this theory is not yet completed today. I will finish it and present it in the near future.

I used only a few shapes for charge carriers in this book. After this works, I simulated electrostatic forces that act on 19 different shape charge carriers. And, I present it to The 2010 Annual Meeting of the Electrostatics Society of America. If you want to read it, please visit ESA home page. It may be published on the related journal.

I limited the material of charge carriers to metal (conductor) only. But, I think asymmetric charged insulator shows different movement against sphere shape charged insulator in electric field.

I hope this book becomes a start point of research on asymmetric charged materials.

Katsuo Sakai
Electrostatic Generator Research Center (Yokohama Japan)
E-mail: gy7a-ski@asahi-net.or.jp
2010/06/09

Chapter 1

ASYMMETRIC FORCE (AF)

1.1. PURPOSE

In a convergent electric field, all bodies are forced to move in the convergent direction by an electrostatic force. This is the gradient force. The other two electrostatic forces (the image force and the Coulomb force) act only on charged bodies, but the gradient force can also act on neutral objects. This is a very useful characteristic. However, a gradient field is not particularly useful. If a non-charged material could be forced to move in a parallel electric field by electrostatic force, this would be a very useful phenomenon.

Thus, the following hypothesis is proposed.

Hypothesis: When a convergent field is changed to a parallel field, if the shape of a non-charged conductor is changed precisely in proportion to the change of the field, the same electrostatic force will act on the altered non-charged conductor.

The above phenomena are proposed to be mathematically equivalent. As a note, in this paper, only the electrostatic force that acts on a conductive body was simulated for simplification.

Therefore, it is a main purpose of this chapter to prove this hypothesis. In addition, finding the optimal shape for this force and confirming this force experimentally are the second and third purposes of this chapter.

1.2. SIMULATION METHODS AND RESULTS

If these two forces that act on the original conductor in a convergent field and the altered conductor in parallel field could be calculated analytically, this task would be easy. However, there is no formula that can be used to calculate these forces for all shapes. There exists only an approximate formula for small spheres in a convergent field [1]. This situation is shown schematically in Figure 1. The following is the approximate formula for a small conductive sphere.

$$Fg = 2\pi r^3 \varepsilon_0 \nabla E_0^{\ 2} \tag{1}$$

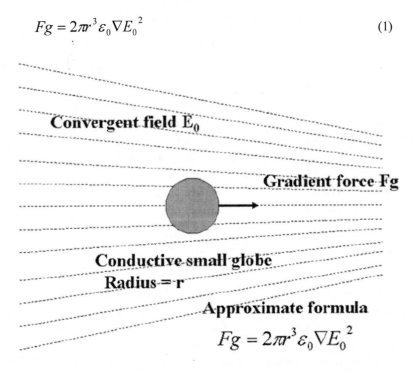

Figure 1. Schematic view of the gradient force and an approximate formula for a sphere.

Therefore, I used a simulation method to calculate these forces. It would be ideal to simulate these forces in three dimensions. However, the simulation program, which was written by the author, is only a bi-dimensional program, employing an axi-symmetric finite difference method. Fortunately, this simulation program can treat an axi-symmetric body. Hence, a cylinder-shaped conductor was selected in place of the conductive small sphere; this cylinder-shaped conductor is hereafter referred to as a cylinder.

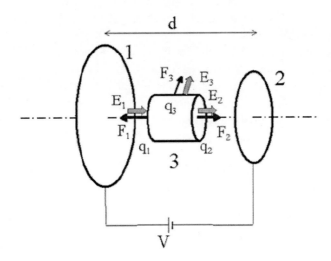

Figure 2. Schematic layout for simulating the electrostatic force acting on a cylinder in a convergent field.

Figure 2 shows a cylinder placed in the convergent field. This field is formed by two disc electrodes that have different diameters.

Figure 2 shows a schematic layout for the simulation. A large circle electrode (**1**), a small circle electrode (2), and a cylinder (3) were aligned along the z axis (the dotted line in Figure 2). The radii of the large electrode, the small electrode, and the cylinder were 250 μm, 100 μm, and 100 μm, respectively. The distance between the large and small electrodes was 600 μm, and the width of the cylinder was 300 μm. A positive voltage of 600 V was applied to the large circle electrode in order to generate a convergent electric field between it and the grounded small circle electrode. The cylinder was electrically floated.

E_1, E_2, and E_3 are the electric field intensities on the left surface, the right surface, and the circumference surface of the cylinder, respectively, while q_1, q_2, and q_3 are the charge quantities of the left surface, the right surface, and the circumference surface of the cylinder, respectively. F_1, F_2, and F_3 are the electrostatic forces acting on the left surface, the right surface, and the circumference surface of the cylinder, respectively.

The total electrostatic force Fe that acts on the conductor was calculated using the following formula, (2)

$$Fe=F_1+F_2 \tag{2}$$

F_3, which acts on the circumference surface of the cylinder, was not included in formula (2) because it cancels out at an interval of 180 degrees and ultimately becomes zero.

The details of this simulation are explained in Appendix 1.

The gradient force that acts on a non-charged cylinder in a convergent field (see Figure 3 (A)) was simulated using an axi-symmetric finite difference method.

Then, the electrostatic force that acts on a cylinder in a parallel field (see Figure 3 (B)) and the force that acts on a piled disk conductor in a parallel field (see Figure 3 (D)) were simulated.

To prove the hypothesis of this paper, first, the radius of the small circle electrode was increased from 100 μm to 250 μm (see Figure 3 (B)). As a result, the electric field between the electrodes changed from asymmetric (convergent) to symmetric (parallel). Next, the radius of the right surface of the cylinder was increased from 100 μm to 200 μm, and the shape of the conductor changed from a symmetric cylinder to an asymmetric bat (see Figure 3 (C)). In this step, the shape of the cylinder must be changed precisely in proportion to the change of the field. However, only 150 μm or 200 μm can be selected in this simulation program; therefore, 200 μm was selected as an approximate value.

As mentioned above, the simulation program, which was written by the author, cannot deal with this bat shape (three dimensions). Hence, the bat shape conductor was replaced with a piled three-disk conductor (see Figure 3 (D)). The radii of the three disks were 100 μm, 150 μm, and 200 μm, and the width of the disks was 100 μm.

Figure 4 shows the design of the cells (mesh) used for simulation (B), and Figure 5 shows the design of the cells (mesh) used for simulation (D). The design of the cells (mesh) used for simulation (A) and an explanation of the cells are given in Appendix 1 (See Figure 49).

Figure 6 shows the simulation results: the field intensity, charge quantity, and electrostatic force of both side surfaces of the cylinder and the piled three-disk conductor under a convergent or parallel field (see Figure 3 (A), (B), and (D)).

For the cylinder placed in the parallel field (see Figure 3 (B)), the field intensity on both surfaces is the same, and consequently, the electrostatic force acting on both surfaces is also the same. For the cylinder placed in the convergent field (see Figure 3 (A)), the field intensity on the right surface is stronger than that on the left surface; as a result, the electrostatic force of the right surface is stronger than that of the left. For the piled disk conductor placed in the parallel field (see Figure 3 (D)), the field intensity on the right surface is weaker than that on the left; nevertheless, the electrostatic force of the right surface is stronger than that of the left because the surface area of the right surface is four times greater than that of the left.

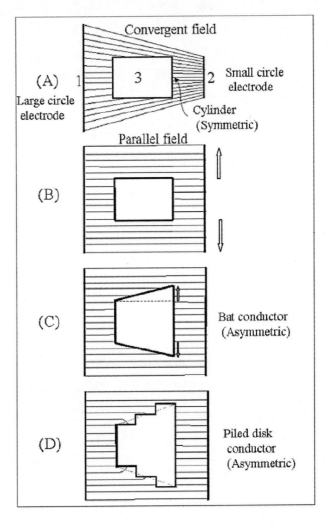

Figure 3. Schematic diagram of electric field patterns and conductor shapes. (A) A convergent field and a cylinder. (B) A parallel field and a cylinder. (C) A parallel field and a baseball-bat-shaped conductor. (D) A parallel field and a piled disk conductor.

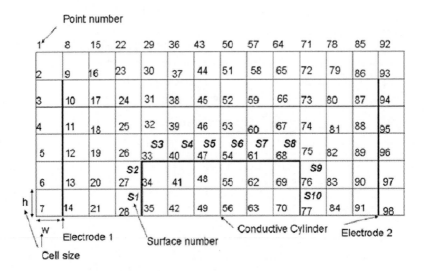

Figure 4. A design of cells (mesh) for simulating the electrostatic force that acts on a non-charged cylinder in a parallel electric field.

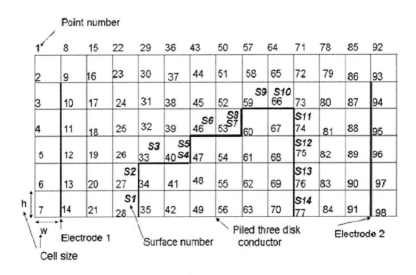

Figure 5. A design of cells (mesh) for simulating the electrostatic force that acts on a non-charged piled three-disk conductor in a parallel electric field.

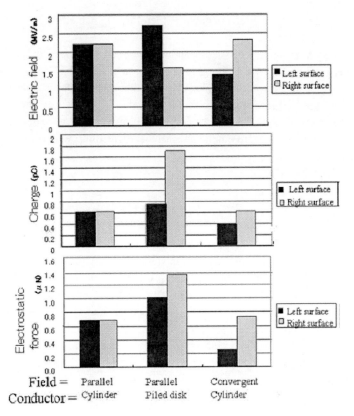

Figure 6. Field intensity, charge quantity, and electrostatic force as a function of field pattern and conductor shape. These variables were simulated for a cylinder or a piled disk conductor in a parallel field or a convergent field.

The electrostatic force of the right surface is stronger than that of the left for the conditions in both Figure 3 (A) and (D), and the differences between the right and left electrostatic forces for both conditions are similar. I propose that this simulation result almost proves the above-mentioned hypothesis.

If the shape of the cylinder had been changed precisely in proportion to the change of the field in Figure 3 (C), and if the electrostatic force acting on this changed cylinder could be simulated, then it would completely agree with the simulated force for Figure 3 (A).

This force (see Figure 3 (A)) is called the gradient force, because the field has a gradient (convergence). Meanwhile, the other force (see Figure 3 (D)) acts on an asymmetric conductor, therefore I named this force the Asymmetric Force (AF).

When a non-charged, small conductive sphere is placed in a gradient (convergent) electric field, the gradient force acts on it. This gradient force can be calculated by formula (1). On the other hand, when a non-charged cylinder is placed in a parallel electric field, the AF acts on it. The formula and the simulation of the AF are completely different in appearance, but I suggest that the two are mathematically equivalent. I will prove this in Appendix 2.

1.3. CAUSE OF THE AF

In the previous chapter, the AF acting on a piled three-disk conductor in a parallel field (see Figure 3 (D)) was simulated. This three-disk conductor has three left surfaces (shown in Figure 3 (D)), where the total area of the three left surfaces is the same as that of the right surface. Nevertheless, the electrostatic force on the right surface is stronger than the total electrostatic force on the combined left surfaces. If a symmetric conductor that has the same area on the left and right surfaces is placed in a parallel field, the electrostatic forces that act on the two surfaces must be the same.

Figure 7 shows the simulation results of the field intensity, charge quantity, and electrostatic force of the first, second, and third left surfaces of the piled conductor.

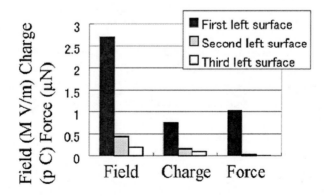

Figure 7. Field intensity, charge quantity, and electrostatic force as a function of the specific surface of the piled three-disk conductor. These are simulated for a piled conductor in a parallel field (see Figure 3 (D)).

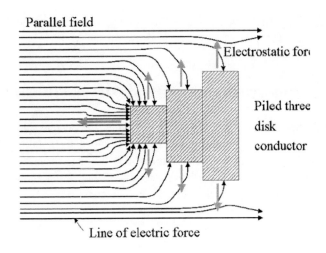

Figure 8. Schematic diagram of the electric force that is concentrated on the asymmetric piled disk conductor in a parallel field.

It can clearly be seen in Figure 7 that the field intensity on the first left surface is much larger than that on the second and third left surfaces; hence, the electrostatic forces that act on the second and third left surfaces are negligible in comparison to that acting on the first left surface. This is because the direction of lines of electric force that runs straight from the left electrode toward the second and third left surface is strongly bent towards the conductor, finally aiming perpendicularly to the circumference surfaces of the first and second disks, as shown in Figure 8.

As a result, the electrostatic force acts on the circumference surfaces perpendicularly. These forces cancel each other at an interval of 180 degrees and ultimately become zero.

This means that the second left surface is electrically shielded by the circumference surface of the first left disk, and the third left surface is electrically shielded by the circumference surface of the second left disk.

On the contrary, the right surface is not shielded. Therefore, many lines of electric force point from the surface to the right electrode. The number of those lines of electric force is the same as the number of lines of electric force that point toward the first left surface and the circumference surfaces of the three disks.

As a result, the electrostatic force acting toward the right becomes larger than the electrostatic force acting toward the left.

Therefore, when an asymmetric conductor consists of surfaces that are perpendicular to the electric field and surfaces that are parallel to the electric field, electrical shielding of a large part of one side (left or right) of the perpendicular surfaces by parallel surfaces causes the AF to arise.

1.4. SIMULATION OF THE BEST SHAPE FOR THE AF

It has been shown that the AF arises when the second and third left vertical surfaces are selectively electrically shielded by parallel surfaces. Accordingly, it would be expected that the AF becomes strongest when the area of the first left vertical surface is minimized with respect to the area of the right vertical surface. Therefore, a suitable shape for this requirement must be a bolt or a cup, as shown in Figure 9.

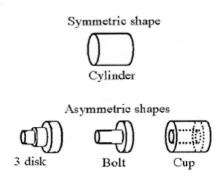

Figure 9. Shapes of conductors used to simulate the AF.

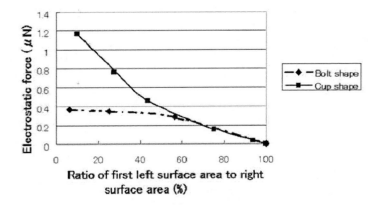

Figure 10. The AF as a function of the ratio of the first left surface area to the right surface area for asymmetric conductors in a parallel field.

To investigate the effect of the ratio of the first left surface area to the right surface area on the AF and to determine the optimum shape of the conductor, numerical simulations were carried out for the following different surface area ratios for the bolt and cup conductors: 5, 10, 20, 30, 40, 50, 70, and 90%.

The simulation results are shown in Figure 10.

Figure 10 shows that the AF of the cup conductor is larger than the AF of the bolt conductor. In particular, when the ratio of the first left surface to the right surface is less than 15%, the AF becomes three times larger than that of the bolt. Therefore, the cup shape is better than the bolt shape; specifically, a thin-walled cup should generate the largest AF.

These results can be explained by the same theory that is given above. The second left surface of the cup (the inner bottom) is perfectly electrically shielded by the surrounding thin wall. In contrast, the second left surface of the bolt cannot be perfectly electrically shielded by the narrow center stick.

Further details of these simulations have been reported in reference [2].

1.5. A SIMPLE EXPERIMENT CONFIRMING THE EXISTENCE OF THE AF

The simulation of the AF was done with the simple structure shown in Figure 2. However, it was difficult to execute a confirmation experiment using this structure. The dimensions of this structure are difficult to achieve experimentally, and the cup shape is very difficult to produce. Therefore, the distance between the electrodes was increased from 0.6 mm to 98 mm. Furthermore, the shape of the conductor was changed from a cup to a box. Figure 11 shows a schematic layout of the experimental setup that was used to confirm the AF.

Before discussing the results of the experiment, I will present a simulation result of the electrostatic force that acts on a box conductor in this situation. This is done because the dimensions of the situation were largely changed. Figure 12 shows the main part of the cell structure used in this simulation. I used a cup in this simulation because my simulation program cannot treat a box, as previously mentioned.

Figure 12 is slightly more complicated in comparison to Figure 49. However, both figures are basically the same. Therefore, an explanation is omitted (an explanation of the cell design is shown in Appendix 1 along with Figure 49).

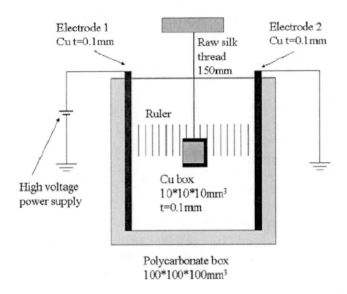

Figure 11. Schematic layout of the experimental setup used to confirm the AF.

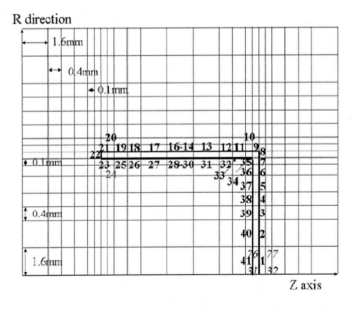

Figure 12. A design of cells (mesh) to simulate the electrostatic force that acts on a charged cup in a parallel electric field.

After running the simulation, the electrostatic force that was found to act on the cup was converted into the electrostatic force that would act on the box, according to the area ratio of the bottom face.

Figure 13 shows the converted electrostatic force as a function of the field intensity between the electrodes. It is apparent that the total electrostatic force increases with the square of the electric field intensity.

Now, I return to Figure 11 for an explanation of the experimental setup and procedure. Initially, a large, transparent, polycarbonate box was prepared. Thin copper plates with dimensions of 100*100 mm^2 and thickness of 0.1 mm were attached to the left and right walls of a large box as left and right electrodes. The left electrode was connected to a high voltage power supply, and the right electrode was grounded.

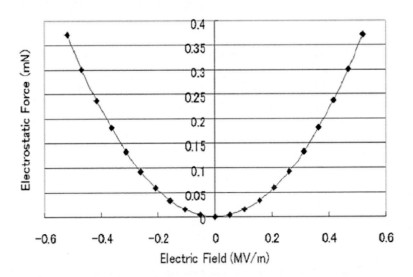

Figure 13. Simulation results of the total electrostatic force acting on a non-charged box in a parallel electric field as a function of the field intensity.

A small left open box conductor (10*10*10 mm^3) that was made from a thin copper plate with a thickness of 0.1 mm was floated by an insulating raw silk thread with a length of 150 mm connected to the center of both electrodes. A ruler was attached to the back wall of the large box to measure the position of the box. A Matsusada Precision Model HAR-50PS was used as a high voltage power supply.

The experiment was performed as follows.

- The left and right electrodes were grounded, and the box was put into contact with one of the grounded electrodes for discharging.
- The box was centered between the grounded electrodes, and the position of the box was measured using the ruler.
- A high voltage was applied to the left electrode; as a result, the box conductor was shifted by a short distance to the right. This altered position of the box was measured using the ruler. The difference between the position measured in step ☐ and the position measured in step ☐ is the shifted distance of the box.
- Steps ①→③ were repeated three times at different voltages (10 kV, 15 kV, 20 kV, 25 kV, 30 kV).
- Finally, the high voltage power supply and the ground line were switched, and the same experiment was repeated.

Figure 14 shows the measured average shifted distance of the non-charged box as a function of the applied voltage.

In Figure 14, it is apparent that the measured shifted distance of the non-charged box increases almost with the square of the applied voltage. This is the same pattern that was observed for the simulation results shown in Figure 13.

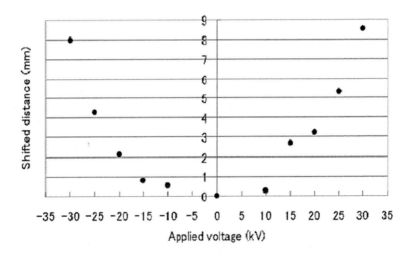

Figure 14. The measured average shifted distance of a non-charged box in a parallel field as a function of the applied voltage.

The shifted distance D of the non-charged box floating in a strong electric field was measured as described above. This shifted distance D is related to the length L and the height Y of the thread as shown in Figure 15.

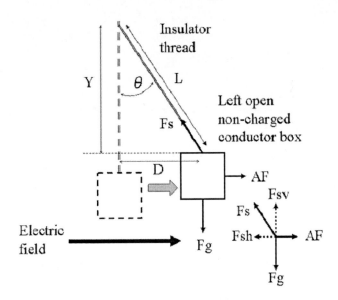

Figure 15. Schematic diagram of the electrostatic force (AF), the gravity force, and the tensile force that act on a non-charged box in a strong parallel field.

The ratio L:D:Y is the same as the ratio of the tensile strength of the thread Fs:the electrostatic force AF:the gravity force Fg. The horizontal component of the tensile force Fsh is equal to the AF, and the vertical component of the tensile force Fsv is equal to the gravity force Fg. Therefore, the AF was calculated from the shifted distance D, the thread length L, and the gravity force Fg, using the following formula (3):

$$AF = Fg \times \tan \vartheta = Fg \times \frac{D}{Y} = Fg \times \frac{D}{\sqrt{L^2 - D^2}} \qquad (3)$$

where L=150 mm, and Fg=4.39 mN, which was calculated from the area of the box, 5 cm^2, the thickness of the Cu plate, 0.1 mm, and the specific gravity of Cu, 8.96.

Figure 16 shows the AF calculated from the measured shifted distance of the box with the simulated AF as a function of the electric field intensity.

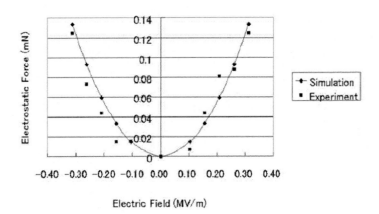

Figure 16. Simulated and experimentally measured AF acting on a non-charged box as a function of the electric field intensity.

As a result, the measured AF shows good agreement with the simulated AF. This result confirms the AF that acts on a non-charged asymmetric conductor in a parallel electric field actually exists.

Further details of this simulation and the experiment have been reported in reference [3].

1.6. APPLICATIONS

Electrostatic transportation of non-charged asymmetric conductors can be realized by using the AF. Traditional electrostatic transportation methods usually require a charged transfer body and a change in potential at the transfer electrodes. In contrast, this new electrostatic transportation method requires neither of these. It needs only an electrostatic field in place of an alternating electric field. The electrostatic field can be easily generated by electrets in place of electrodes. As a result, the structure of this new electrostatic transportation machine becomes very simple.

Electrostatic acceleration of non-charged asymmetric conductors can be realized by using the AF as well. Traditional electrostatic accelerators usually require a charged target body and an increased potential of the accelerative electrodes that is generated by high voltage power supplies. In contrast, this new electrostatic accelerator does not require either of these; it needs only a constant electrostatic field in place of an increasing electrostatic field. The constant electrostatic field can be easily generated by electrets. As a result, the structure of this new electrostatic accelerator machine becomes very simple.

A new type of electrostatic motor can also be realized by using the AF. The traditional electric motor consists of moving electrodes and fixed electrodes. The moving electrodes are given charges, and the fixed electrodes are applied to an alternating voltage. In contrast, this new electrostatic motor requires neither; the moving electrodes are not given charges, and the fixed electrodes are applied to a constant voltage. The constant voltage can be easily generated by electrets. As a result, the structure of the new electrostatic motor becomes very simple.

1.7. CONCLUSION

1) The following hypothesis has been almost proven by simulation.

Hypothesis: When a convergent field is changed to a parallel field, if the shape of the non-charged conductor is changed precisely in proportion to the change of the field, the same electrostatic force will act on the changed conductor.

Thus, this force is termed the AF

2) It is thought that the AF is caused by surfaces that are perpendicular to the electric field that are also electrically shielded along with surfaces that are parallel to the electric field.

3) The AF is the strongest when the conductor is cup-shaped.

4) The existence of the AF has been confirmed by a simple experiment.

5) Electrostatic transporters, electrostatic accelerators, and electrostatic motors can be greatly improved by using the AF.

Chapter 2

THE ASYMMETRIC IMAGE FORCE (AIF)

2.1. PURPOSE

When a charged body is placed near an electrode, an electrostatic force acts on this body. This force is known as the image force. If the body is a spherical conductor, the force is calculated by the following formula (4) [4].

$$f = \frac{q^2}{4\pi\varepsilon_0 (2x)^2} \tag{4}$$

where q: Quantity of charge on the sphere.

r: Distance between the sphere's center and the grounded electrode.

ε_0: Vacuum permittivity

This force does not change even if the sphere is turned in either direction because the shape is symmetric. In contrast, if an asymmetric charged conductor is turned to the right or left, this force may change.

Therefore, confirming this expectation is the purpose of this chapter.

2.2. SIMULATION RESULT

Figure 17 shows the cell layout (mesh) of the simulation of the image force that acts on a charged cup.

In this step, the electrostatic force that acts on a charged cup near a grounded electrode was simulated. The height, radius, and thickness of the cup are 10 mm, 5 mm, and 1 mm, respectively.

The quantity of charge on this cup is +2.0 nC. The distance between the grounded electrode and the cup varied from 1.0 mm to 9.0 mm. In the first simulation, the mouth of the cup was pointed toward the grounded electrode. In the second simulation, the bottom of the cup was pointed toward the grounded electrode, as shown in Figure 17.

The conventional image force that acts on a spherical conductor near a grounded electrode was calculated according to formula (4). The radius of the sphere is 5 mm, and the quantity of charge on the sphere is +2.0 nC.

Figure 18 shows the simulated image forces and the calculated image force.

From Figure 18, it is apparent that when the bottom of the cup is pointed toward the grounded electrode, the simulated image force is large, and when the mouth of the cup is pointed toward the grounded electrode, the simulated image force is small.

I called this changeable image force the Asymmetric Image Force (AIF).

(Caution: when the distance is 1 mm, the calculated image force becomes smaller than the simulated force. This is not a real phenomenon. It is well known in toner development research that the real image force is about two times bigger than the image force calculated by formula (4) when a toner is placed on an electrode [5].)

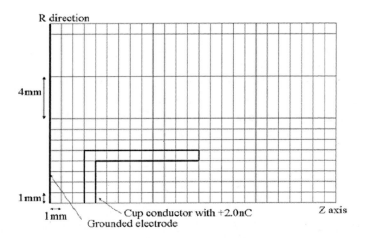

Figure 17. A design of cells (mesh) for simulating the image force that acts on a charged cup near a grounded electrode.

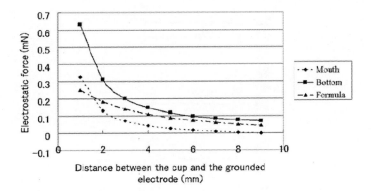

Figure 18. Simulation results for the image force acting on a charged cup and the calculated image force acting on a charged sphere near a grounded electrode.

2.3. EXPERIMENTAL RESULTS

I attempted to confirm these simulation results using an experimental setup that is explained in the next chapter. However, the experiment failed because the force of gravity on the charged box in air did not balance the electric force near the electrode. the charged box always came into contact with the electrode when it was made to approach the electrode.

Thus, another instrument is needed to measure the AIF.

2.4. CAUSE OF THE AIF

It is proposed that this force is caused by the charge distribution rather than electric shielding, because there are charges on the cup, but there is not an electric field surrounding the cup.

Figure 19 shows the simulated charge distribution of the cup.

It is apparent in Figure 19 that about half of the charges gather on the first left surface (outer bottom) when the bottom is pointed toward the grounded electrode, but only 20% of the charges gather on the first left surface (edge of the mouth) when the mouth is pointed toward the grounded electrode, as shown in Figure 19.

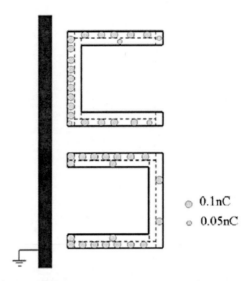

Figure 19. Charge distribution in a cup placed near a grounded electrode. The bottom of the upper cup is pointed toward the electrode, and the mouth of the lower cup is pointed toward the electrode. Both cups have a charge of +2.0 nC each.

This decrease in the number of charges gathering on the first left surface occurs because the area of the first left surface in the former situation is large, and the area of the first left surface in the latter situation is very small. As a result, a large electrostatic force acts on the bottom in the former situation, but a weak electrostatic force acts on the edge in the latter situation.

Many of the other charges are trapped on the circumference surface of the cup. The electrostatic force that acts on those charges is perpendicular to the circumference surface. These forces cancel each other at an interval of 180 degrees, thus becoming zero.

Therefore, the cause of the AIF is the following.

The charged conductor has a small first left (or right) vertical surface, a large first right (or left) vertical surface, and a horizontal surface. When the large surface is pointed toward a grounded electrode, many charges can gather at the large surface. In contrast, when the small surface is pointed toward the grounded electrode, only a small amount of charge can gather at the small surface. The other charges are stopped at the horizontal surface, and the electrostatic force that acts on those charges does not contribute to the overall electrostatic force because it cancels out at an interval of 180 degrees.

2.5. APPLICATIONS

At this point, I cannot present any electrostatic applications that would use the AIF.

2.6. CONCLUSION

The following expectation is confirmed by a simulation. If the direction of an asymmetric charged conductor near a grounded electrode is turned to the right or left, the magnitude of the image force on the conductor should change.

Chapter 3

THE ASYMMETRIC COULOMB FORCE

3.1. PURPOSE

When a point charge is placed in an electric field, an electrostatic force acts on this charge. This force is known as the Coulomb force and can be calculated by the formula (5).

$$f = qE \tag{5}$$

where: q: Quantity of the charge

E: Intensity of the field

This formula is used not only for point charges, but also small charged conductive spheres. In this case, the absolute value of this force does not change when the direction of the electric field is reversed. However, it is expected that if the shape of the charge carrier is asymmetric, then the absolute value of the force will change when the direction of the electric field is reversed, in analogy with the AF and AIF.

I named this changeable Coulomb force the Asymmetric Coulomb Force (ACF). Therefore, confirming ACF by a simulation and an experiment is the first purpose of this chapter. The second purpose of this chapter is to clarify the cause of the ACF.

3.2. SIMULATION RESULTS

ACF is similar to AF. The box (cup) conductor is neutral in the AF case. In contrast, the box (cup) conductor is charged in the ACF case. Therefore, the experimental setup for confirming the ACF is the same as that used for the AF (see Figure 11). Thus, the cell layout for the ACF simulation is also the same as the cell layout for the AF simulation. Hence, I can use the cell layout of the AF (see Figure 12) as the cell layout of the ACF. However, I changed the radius of the cup from 5.0 mm to 5.6 mm. As a result, the area of the bottom surface of the cup is the same as the area of the bottom surface of the box that was used in the following experiment. Figure 20 shows the new cell layout used to simulate the ACF.

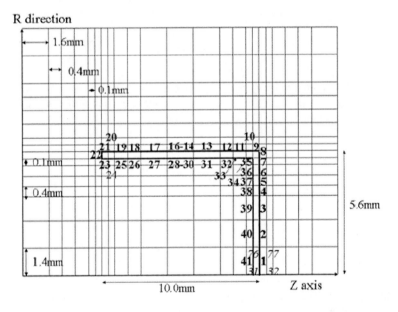

Figure 20. A design of cells (mesh) to simulate the electrostatic force that acts on a charged cup in a parallel electric field.

Because the simulation method (the axi-symmetric finite difference method) of the ACF is the same as that of the AF, the explanation is shortened here. I explain the simulation procedure only with Figure 21.

In the corresponding experiment that will be explained later, the box is charged by electrostatic induction. Therefore, the quantity of the induction charge is simulated first. When electrode 1 is grounded, a voltage of 10 kV is applied to electrode 2, and the cup is brought into contact with the grounded electrode (see Figure 21), the induction charge on the cup is simulated as -1.026 nC.

Therefore, in this simulation, the cup conductor was initially charged to −1.026 nC, then electrode 1 was grounded and electrode 2 was set to a high voltage (from +10 kV to +30 kV). I defined this condition as a forward electric field for the negative charge (see Figure 21); in this case, the electrostatic force pulled the charged cup toward the right side.

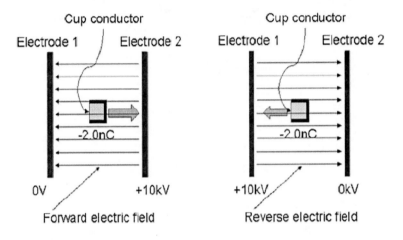

Figure 21. Explanation of the forward electric field and the reverse electric field.

Next, the cup was charged to −2.052 nC, and third, the cup was charged to −3.078 nC, and the electrostatic forces that act on the charged cup were simulated. Finally, electrode 1 was set at a high voltage and electrode 2 was grounded. I defined this condition as a reverse electric field for the negative charge (see Figure 21); in this case, the electrostatic force pulled the charged cup toward the left side.

Figure 22 shows the simulated electrostatic force in the forward electric field and the reverse electric field scenarios as a function of the field intensity between the electrodes.

In Figure 22, it is apparent that when the absolute value of the forward electric field intensity is equal to that of the reverse electric field, the electrostatic force that acts on the charged cup in the forward field is larger than the electrostatic force in the reverse field. It is also apparent that when the electric field becomes strong, the electrostatic force generally increases.

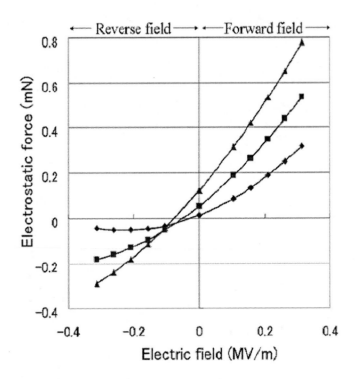

Figure 22. Simulation results for the electrostatic force acting on a negatively charged cup in a forward electric field and a reverse electric field.

(In the case that the quantity of the charge is small (-1.026 nC) and the reverse electric field is strong (-0.3 MV/m), the electrostatic force becomes small. This result can be explained by AF. In the case in which the electric field is 0.0 MV/m, the electrostatic force does not become zero.

This result can be explained by the AIF. If the distance between the cup and the right-side electrode is large enough, this unexpected result no longer occurs.)

3.3. EXPERIMENT RESULTS

Because the experiment equipment for the ACF is the same as that of the AF (Figure 11), the explanation is shortened here. I explain the experiment procedure only with Figure 23.

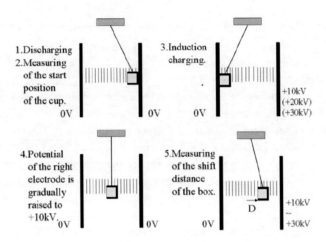

Figure 23. Schematic of the procedure of the experiment to confirm the ACF.

This experiment was performed using the following procedure:

- The left and right electrodes were grounded, and the box conductor was brought into contact the grounded electrode for discharging.
- The non-charged box conductor was centered between the electrodes, and the position of this box was measured using the ruler.
- The right electrode was charged to V1 (+10 kV), and the box conductor was brought into contact with the grounded left electrode for induction charging. It was then kept away from the electrode so that a negative charge was maintained on the box conductor.
- The right electrode was grounded again, and it was gradually charged to the target high voltage V2 (+10 kV), creating a forward electric field for negative charges between the electrodes. As a result, the box conductor shifted toward the right side. This position of the box was measured using the ruler. The difference between the position measured in step ④ and the position measured in step ② was the shifted distance of the box.
- ①→ ④ This procedure was repeated with varying values of V1 (+20 kV, +30 kV) and V2 (+15 kV, +20 kV, +25 kV, +30 kV).

Finally, the left electrode was charged to a positive high voltage and the right electrode was grounded, creating a reverse electric field for negative charges between the electrodes, and the same experiment was performed again.

The following three photographs show the typical three positions of the left open box of the experiment.

Figure. 24 shows the measured shifted distance of the box as a function of the applied high voltage.

It is apparent from in Figure 24 that the measured shifted distance of the box in a forward electric field is large, but it is small in a reverse electric field. FurthermoreAnd, we can recognize see that the pattern inof Figure 24 (experiment) is about same as patternsimilar to that of Figure 22 (simulation).

HoweverBut, for the correct comparingto appropriately compare between Figure 22 (simulation) and Figure 24 (experiment), the shifted distance of the box must be changed converted to the electrostatic force.

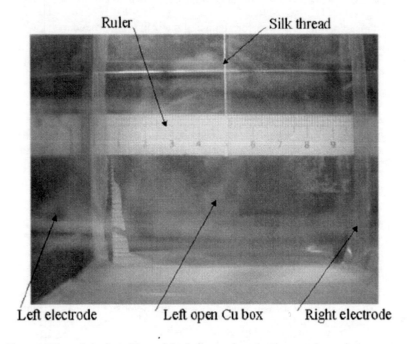

Photo 1. The original position of the left open box that has no charge in no electric field.

Photo 2. The shifted position of the left open box that has charge of -2.0nC in forward electric field of +0.2MV/m.

Photo 3. The shifted position of the left open box that has charge of -2.0nC in reverse electric field of -0.2MV/m.

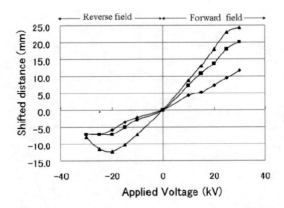

Figure 24. The measured shifted distance of the box in the forward electric field and the reverse electric field as function of the applied voltage.

3.4. A COMPARISON OF THE EXPERIMENTAL AND SIMULATION RESULTS

The shifted distance D of the charged box conductor floating in a strong electric field was measured as described above. This shifted distance D is related to the length L and the height Y of the thread, as shown in Figure 25.

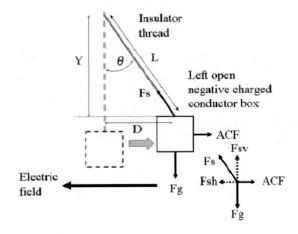

Figure 25. Schematic diagram of the electrostatic force (ACF), the gravity force, and the tensile force acting on a charged box in a strong electric field.

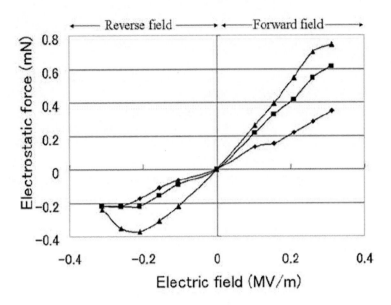

Figure 26. Experimentally measured ACF acting on a charged box as a function of the electric field intensity.

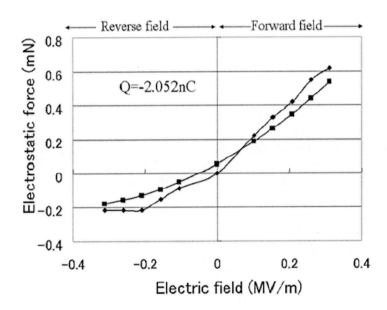

Figure 27. Experimentally measured and simulated ACF acting on a charged box as a function of the electric field intensity. The charge quantity is –2.052 nC.

In Chapters 1-5, I explained how to calculate the electrostatic force that acts on the box from the shifted distance. Therefore, the explanation is omitted here.

The experimentally measured ACF acting on the charged box in a strong electric field is shown in Figure 26.

The overall pattern of the experimental result is very similar to that of the simulation (see Figure 22). However, it is difficult to compare the results of the experiment and the results of the simulation when the three simulation results and the three experiment results are shown simultaneously on one graph. Thus, Figure 27 shows the results obtained for a charge of -2.052 nC. The other results can be confirmed in reference [6].

It is apparent in Figure 27 that the experimental result shows good agreement with the simulation result as a general rule and that ACF due to the forward electric field is larger (by a factor of 2 or more) than ACF due to the reverse electric field. Therefore, the ACF has almost been confirmed via simulation and experimentation.

3.5. CAUSE OF THE ACF

As mentioned above, the existence of the ACF has been confirmed experimentally. Therefore, I next focus on the cause of the ACF.

For this purpose, a cup-shaped conductor was divided into two parts, as shown in Figure 28. The first part is a pipe, and the other is a disk. The electrostatic forces that act on these three conductors were simulated using the same simulation method described previously. In this simulation, the same charge of -2.06 nC was placed on the three conductors. The cell layouts of the pipe and the disc were omitted, because they can be easily determined from Figure 20.

Figure 29 shows the simulated electrostatic forces of the three conductors in the forward electric field and the reverse electric field as a function of the field intensity between the electrodes. The Coulomb force calculated by formula (5) is also shown in Figure 29. For this calculation, the charge q was fixed at -2.052 nC.

In Figure 29, the triangle data points show the electrostatic force that acts on the cup, the square data points show the electrostatic force that acts on the pipe, and the diamond data points show the electrostatic force that acts on the disk. The X data points show the result calculated from formula (5).

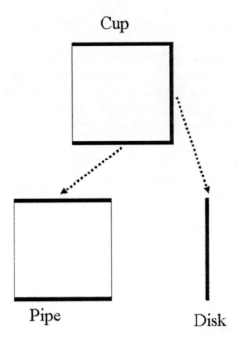

Figure 28. Original cup conductor and the cup divided into two parts (pipe and disk).

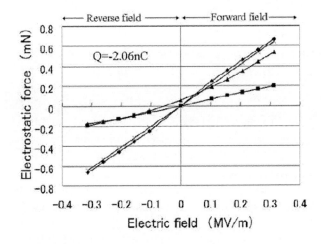

Figure 29. The simulated electrostatic forces that act on the three differently shaped conductors and the calculated Coulomb force.

The following results are apparent from Figure 29.

(1) The simulation results of the disk and the cylinder are point symmetric. Both shapes are symmetric. The calculation result of the Coulomb force is point symmetric as well. This means that the absolute value of the electrostatic force in the forward field is equal to that in the reverse field.

(2) The simulated electrostatic force of the disk shows good agreement with the calculated Coulomb force. It is not presently clear whether this consensus occurred by chance.

(3) The electrostatic force of the disk is about three times stronger than the electrostatic force of the pipe.

(4) The simulation result of the cup is not point symmetric. In the forward field, the electrostatic force of the cup is similar to the electrostatic force of the disk.

In the reverse field, the electrostatic force of the cup is equal to the electrostatic force of the pipe.

As a result, the electrostatic force of the cup in the forward field is about three times stronger than the electrostatic force of the cup in the reverse field.

These results suggest that the charge distribution of the cup is similar to the charge distribution of the disk in the forward field, and the charge distribution of the cup is similar to the charge distribution of the pipe in the reverse field.

Therefore, I simulated the charge distribution of the three conductors in both electric fields. Figure 30 schematically shows the simulation results.

In Figure 30, the black circles represent electrons, and the blank circles represent holes. The large circles and small circles represent 0.1 and 0.05 nC, respectively. The arrow marks represent the direction and intensity of the electric field. The large and small arrows represent 0.5 MV/m and 0.25 MV/m, respectively.

(Caution: In Figure 30, the circles and arrows are divided on the left and right with different numbers for convenience. For example, the left edge of the pipe has four arrows, and the right edge of the pipe has three arrows in the reverse electric field. However, this does not mean that the intensity of the electric field on the left edge is 2.0 MV/m and the intensity of the right edge is 1.5 MV/m. The correct meaning is that the intensity of the edge is 3.5 MV/m.)

(Caution: There are a few holes in Figure 30. They were generated by electrostatic induction.)

It is apparent in Figure 30 that the charge distribution of the cup is the same as the charge distribution of the pipe in the reverse field. However, the charge distribution of the cup is not the same as the charge distribution of the disk in the forward field. Therefore, this phenomenon cannot be explained solely by the charge distribution.

It is apparent in Figure 30 that the outer bottom of the cup has a large amount of charge in the forward field. On the other hand, the edge of the mouth of the cup does not have a large charge. As a result, a strong forward electrostatic force acts on the cup in the forward field.

Of course, the outer circumference surface is charged as well, but the electrostatic force that acts on those charges is perpendicular to the circumference surface and cancels out at an interval of 180 degrees.

In contrast, when the cup is placed in the reverse electric field, all of the electrons leave the outer bottom of the cup, but none of the electrons stop at the inner bottom of the cup because this place is electrically shielded by the surrounding cup wall. If this region was not shielded, many electrons would stop here and a strong electrostatic force in the reverse direction would be generated.

Because none of the electrons stop at the inner bottom, all of the electrons move toward the mouth of the cup. A few electrons migrate to the edge, but many electrons stop at the circumference surface of the cup.

This occurs because the area of the edge is too small for many electrons to gather. Electrons repulse each other with electrostatic force; thus, only a few electrons can remain in the small area.

As mentioned above, the electrons on the circumference surface make no contribution to the effective electrostatic force. Finally, only the electrostatic force that acts on the edge remains. However, this force is weak, Of course, the intensity of the electric field on the edge is strong, but the quantity of electrons on the edge is small. Therefore, a weak reverse electrostatic force acts on the cup in the reverse field.

Therefore, the cause of the ACF is summarized in the following:

When a charged conductor consists of four parts, that is a small first left surface, a large second left surface, a large right surface, and a circumference surface, and the left and right surfaces are perpendicular to the electric field and the circumference surface is parallel to the electric field, in the reverse electric field, the large second left surface is electrically shielded by the circumference surface and only a few charges can gather at the non-shielded small first left surface. The other charges that remain on the circumference surface cannot generate an effective electrostatic force. In contrast, in the forward electric field, the large right surface is not shielded, and many charges can gather at the right surface.

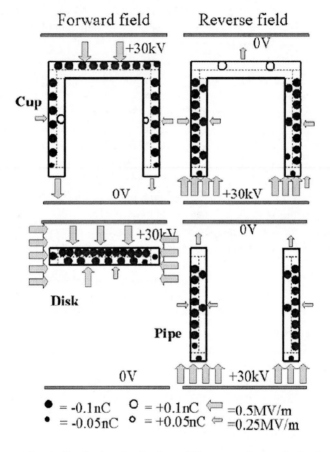

Figure 30. Charge distribution for the three different conductors in the forward and reverse fields.

3.6. CONCLUSION

(1) The existence of the ACF that acts on a charged asymmetric conductor in a parallel electric field has been confirmed by a simulation and an experiment.

(2) The cause of the ACF has been clarified. Selective electric shielding and an unbalanced charge distribution are the main reasons behind this force.

Chapter 4

A NEW ELECTROSTATIC GENERATOR (AN APPLICATION OF THE ACF)

4.1. PURPOSE

As mentioned in the abstract, a new electrostatic generator can be designed to use the ACF. Therefore, it is the first purpose of this chapter to clarify the difference between the traditional generator and the new generator. A determination of the expected performance of the new generator is the second purpose. The concrete design and manufacturing method of the new generator comprise the third purpose. A prediction of the performance of this new generator is the fourth purpose.

4.2. BASIC THEORY

The idea behind an electrostatic generator has been defined by lifting the charge to a high potential by mechanical force against the electric force that acts on this charge. It is impossible for the mechanical force to carry the charge directly. Therefore, the charge is placed on a suitable body. We call this body the charge carrier.

The most popular electrostatic generator is the Van de Graaff type electrostatic generator. This was invented by Dr. Van de Graaff in 1931 in the USA. Today, it is used with a large voltage power supply. It can produce ten million volts. In this machine, an insulating belt is used as a charge carrier. Figure 31 shows an example of this generator.

The insulating belt is moved in the direction of the arrow by a motor. The bottom corona discharge pin array places positive ions on the insulating belt. The positive ions on the insulating belt are carried to the high voltage electrode sphere by the mechanical force of a motor.

Corona discharge occurs between the negative charge on the recovery pin array and the positive ions on the insulating belt. As a result, the positive ions on the insulating belt are neutralized by the negative corona ions. Then, positive charges (holes) are added to the high voltage electrode sphere.

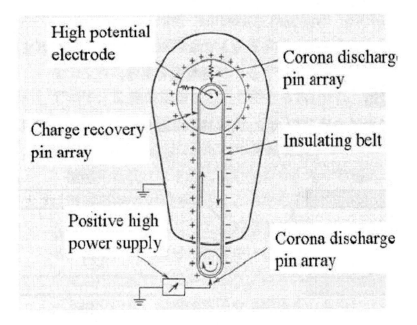

Figure 31. Schematic layout of the Van de Graaff electrostatic generator [7].

The principle of this electrostatic generator is shown schematically in Figure 32.

In Figure 32, the bold line represents the potential, and the arrows represent the forces. The small black circles represent the electrons, and the gray plates represent the charge carriers.

In the Van de Graaff electrostatic generator, the charge carrier is directly transported by a strong mechanical force, Fm, against the electrostatic force Fe.

In contrast, in the new electrostatic generator, the charge carrier is firstly moved in the forward electric field caused by electrets according to the electrostatic force Fe1. In this process, the charge carrier gains kinetic energy from the electric field. Then, the carrier is moved in the reverse electric field, expending the given energy against electrostatic force Fe2.

The shape of this charge carrier is asymmetric.

Therefore, the ACF acts on this charge carrier. Thus, the absolute value of Fe1 is larger than that of Fe2. As a result, the charge carrier can arrive at a potential that is larger (-200 V) than the initial potential (0 V).

The new electrostatic generator cannot produce ten million volts, but it does not require mechanical force. If the lifetime of the electret was infinite, the new electrostatic generator could generate electric energy forever without adding energy.

As a result, this new electrostatic generator can solve the CO_2 problem and the energy crisis at the same time.

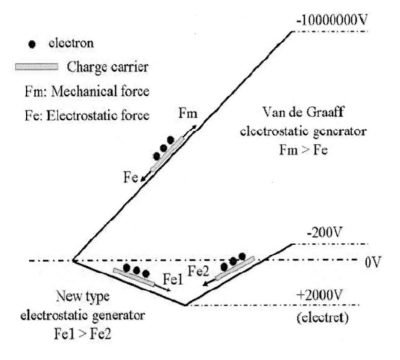

Figure 32. Schematic explanation of the principles behind the two electrostatic generators.

4.3. SIMULATION RESULTS OF THE ENERGY GAINED USING THE NEW ELECTROSTATIC GENERATOR

Now that the basic principle of the new electrostatic generator is clear, I show the smallest unit of the new generator concretely and simulate the energy gained by the smallest unit. Figure 33 shows the smallest unit of the new generator schematically.

This generator mainly consists of electrets 2, induction electrodes 1, and recovery electrodes 4. All of them are disposed on insulating base board 5. The distance between the boards is 0.240 mm. The width of the induction electrodes, the electrets, and the recovery electrodes is 0.080 mm, 0.120 mm, and 0.160mm respectively. The electrets, the induction electrodes, and the recovery electrodes have the same length of 38.000 mm. The distance between the electrets and the induction electrodes is 0.320 mm, and the distance between the electrets and the recovery electrodes is 0.320 mm as well.

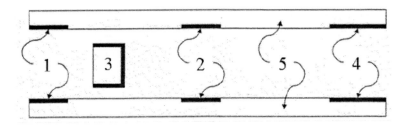

Figure 33. Schematic layout of one unit of the new electrostatic generator.

The electrets have a semi-permanent positive charge density of +0.1 mC/m^2. The induction electrodes are grounded. The recovery electrodes are kept at a voltage of -260 V.   As a result, the electric potential of the electrets is approximately +2600 V.   Therefore, the electrets and the induction electrodes produce a forward electric field for a negative charge between them.

The electrets and the recovery electrodes produce a reverse electric field for a negative charge between them.

The long open box 3 is used for a charge carrier that carries negative charge (electron) from the induction electrodes to the recovery electrodes through the electrets. The height of the long box is 0.160 mm, and the width of the long box is 0.08 mm. The thickness of the wall of the long box is 0.02 mm. The length of the long box is 38.000 mm.

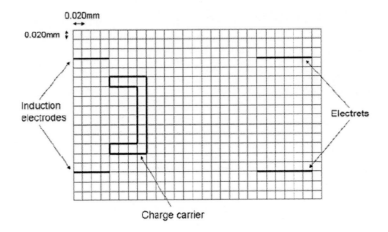

Figure 34. A design of cells (mesh) to simulate the electrostatic force that acts on a charge carrier in one unit of the new electrostatic generator.

Figure 34 shows the left half of the cell design (mesh) for the simulation of one unit of the new electrostatic generator. As mentioned previously, I used the axi-symmetric finite difference method to simulate the electrostatic force that acts on a cup. This time, I must simulate the electrostatic force that acts on a long box; therefore, I used the normal (XY) finite difference method. Both methods are basically equivalent; therefore, a detailed explanation is not given.

At first, the quantity of induction charge to the charge carrier from the induction electrode was simulated. Figure 34 shows the position of the charge carrier at which the induction charge was simulated. At this location, the potential of the charge carrier was fixed to zero volts. The total surface charge of the charge carrier was calculated according to formula (13) (see Appendix 1).

The resulting charge was 0.586 nC. Therefore, this charge carrier was given 0.586 nC in this simulation. The ACF that acts on this conductor was simulated using formula (2) at an interval of 0.020 mm. The recovery electrodes were grounded in this simulation for simplicity. Figure 35 shows the results of the simulation.

The vertical axis shows the ACF, and the horizontal axis shows the distance between the induction electrode and the charge carrier (left side). The ACF in the forward field is positive; therefore, the charge carrier gains kinetic energy. This energy w was calculated by formula (6).

$$w = \sum F * s \ \text{[J]} \tag{6}$$

where: F=ACF [N]
    s=20*E-6 m
    The resulting energy was 5.70E-7 J.

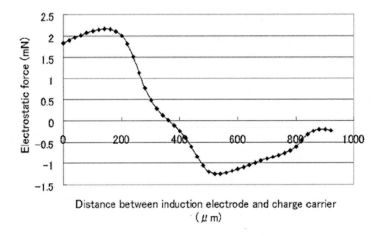

Figure 35. ACF vs. distance between the charge carrier and the induction electrode.

In contrast, the ACF becomes negative in the reverse electric field. Then, the charge carrier loses some of its kinetic energy. The energy loss is calculated as 4.17E-7 J. As a result, the charge carrier maintains some kinetic energy, 1.53E-7 J, when it arrives between the recovery electrodes. The carried charge quantity Q is -5.86E-10 C. This charge Q can be lifted to a higher potential by the extra energy w. This possible potential V is calculated by formula (7).

$$V = \frac{w}{Q} \ \text{[V]} \tag{7}$$

The resulting potential is 261 V.

The carried charge cannot be recovered perfectly, because the recovery electrodes do not comprise a perfect Faraday gauge. The remaining charge on the charge carrier can be simulated by grounding the charge carrier between the recovery electrodes. The simulated remaining charge was -

0.61E-10 C. As a result, the recovered charge was -5.25E-10 C, and the recovery rate was 90%.

In this one process, a charge of -5.25E-10 C was lifted to -261 V from 0 V.

The generated electric energy, We, can be calculated by formula (8).

$$We = Q * V \ \text{[J]} \tag{8}$$

The resulting energy was 1.37E-7 J. This is a small amount of energy, but a large amount of energy can be obtained by gathering those small units. The concrete design of the new electrostatic generator that can generate a large amount of energy is explained in the following chapter.

4.4. MANUFACTURING METHOD OF A NEW ELECTROSTATIC GENERATOR

Figure 36 shows a very simple structure of the box-type electrode that is used as the charge carrier.

This simple electrode 3 can easily be made as follows.

(1) A copper foil is placed on insulating base **6.**
(2) The unwanted part of the copper foil is removed by etching.

This is a very simple method, but the use of space is inefficient, because the thickness of the base is large and the thickness of the electrode is small.

Figure 37 shows an ideal structure of the electrode that is used as the charge carrier.

In this ideal charge carrier disc, many long box electrodes 3 are oriented radially to the circumference 62 from the core 61 of the insulating base disc. This charge carrier disc is sandwiched between two electrode discs. Figure 38 shows one set of three disks of the new electrostatic generator.

In Figure 38, mark 6 shows the charge carrier disc. Marks 51 and 52 show the upper electrode disc and the lower electrode disc, respectively. Mark 3 shows the charge carrier electrode. Marks 1, 2, and 4 show the induction electrode, the electrets, and the recovery electrode, respectively. These electrodes and electrets are made on one side of the electrode disc.

The lower electrode disc has the same structure as the upper electrode disc. However, the electrode surfaces of the two electrode discs face each

other. These two electrode discs are fixed, and the charge carrier disc rotates
at a high speed.

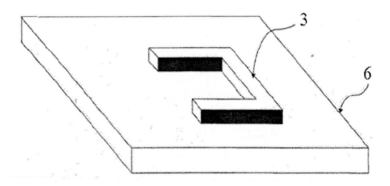

Figure 36. Simple structure of the box-type electrode that is used as the charge
carrier.

The three discs have the same radius of 60 mm. The three electrodes and
the electret have the same length of 38 mm. They are oriented radially from
20 mm to 58 mm from the center of the disk. Figure 39 shows the concrete
design of one unit of the new electrostatic generator. The measurements
given in Figure 39 are the measurements of a part displayed by an alternating
long and short dashed line in Figure 38, and the units are millimeters.

The marks (1, 2, 3, 4, 51, 52)used in Figure 39 correspond to those in
Figure 38. The measurements of each part are the same as for the above-
mentioned simulation. However, one pair of induction electrodes was added.

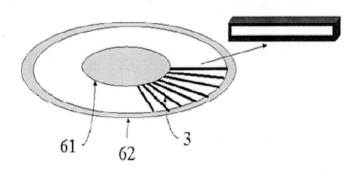

Figure 37. Ideal structure of the box-type electrodes to be used as charge carriers.

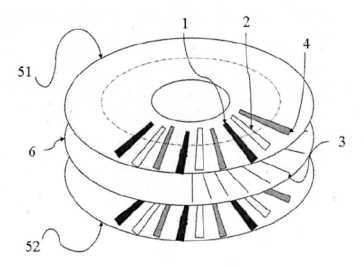

Figure 38. Schematic layout of the one set of three disks of the new electrostatic generator.

They are drawn at the right side in Figure 39. The distance between the recovery electrode and the added induction electrode is 0.20 mm. As a result, the width of one unit of the new electrostatic generator is 1.32 mm. In addition, two more charge carriers were added. As for the die center distance, 0.44 mm was selected to avoid interference.

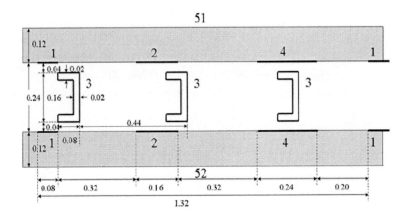

Figure 39. A detailed layout of one unit of the new electrostatic generator.

The induction electrodes and the recovery electrodes can be easily made by the above-mentioned etching method. The electrets can be made as follows.

(1) The base plate (a polycarbonate disc) is warmed to 100-120 degrees.
(2) The unwanted part of this disc is covered by a mask.
(3) This covered disc is exposed to a corona discharge.
(4) Then, this disc is rapidly cooled to the freezing point.

As a result, corona ions are trapped on the polycarbonate surface, which becomes an electret. Figure 40 shows the manufacturing method of the charge carrier disc. The charge carrier disc can be made by an improved etching method. It is made through the following process.

(1) A polycarbonate disk with a thickness of 0.12 mm is cut as shown in Figure 40 (1) to make the slits. The length and the width of the slits are 38 mm and 0.22 mm, respectively. The die center distance of slits is 0.44 mm. An enlarged image of the small area of the center of the slits, displayed by an alternating long and short dashed line in Figure 40 (1) right, is shown at Figure 40 (1) left.
(2) The slits are filled with sacrifice resin. Mark 62 shows the filled slits, and mark 61 shows the polycarbonate base.
(3) Copper foils (t=0.02 mm) are placed on both sides of the disc. Mark 30 shows the copper foils.
(4) The unwanted part of the copper foil is removed by etching.
(5) The polycarbonate base between the sacrifice resin is cut.
(6) The walls of the right side of the sacrifice resin are plated with copper when a voltage is applied to both the top and bottom copper foils. However, the walls of the left side of the sacrifice resin are not plated, because the electric field is weak. Mark 31 shows the plated copper.
(7) The sacrifice resin is dissolved in and is removed.

Now, we can manufacture the new electrostatic generator that is shown in Figure 38. In the next section, I will predict the performance of the new electrostatic generator.

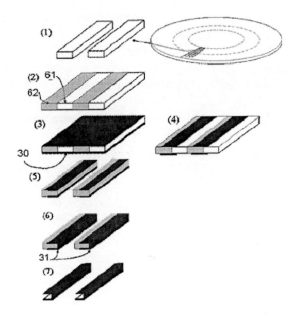

Figure 40. Manufacturing method of the charge carrier disc of the new electrostatic generator.

4.5. PREDICTED PERFORMANCE OF THE NEW ELECTROSTATIC GENERATOR

As mentioned before, the gained electric energy was simulated as 1.37E-7 J, when the charge carrier arrived at the recovery electrode. Therefore, the generated electric power P of one set of three discs can be calculated by formula (9).

$$P=n*m*1.37E-7 \text{ W} \tag{9}$$

where n: Number of charge carriers that arrive at one recovery electrode in 1 sec.

m: Number of recovery electrodes on one electrode disc.

The value of n was calculated to be 9280 according to the pitch and velocity of the charge carrier. As mentioned above, the pitch is 0.44 mm, and the velocity is 4.083 m/sec when the rotational speed of the charge carrier disc is 1000 rpm.

The number m of the recovery electrodes on the electrode disk was calculated to be 185 according to the pitch of the unit and the circumference length of the charge carrier. As mentioned above, the pitch is 1.32 mm and the circumference length is 245.0 mm at the center of the recovery electrodes, which is displayed by an alternating long and short dashed line in Figure 38.

Here, n=9280 and m=185 are substituted into expression (9), and P is 0.236 W.

This is the electric power generated by one set of three discs of the new electrostatic generator shown in Figure 38. The thickness of one set is 0.480 mm, as mentioned before. However, the thickness can be reduced to 0.360 mm by using a double-electrode disc. The structure of this disk is drawn schematically in Figure 41.

The marks in Figure 41 correspond to those in Figure 39. The thickness of the electrode disc and the charge carrier disc combined is 0.360mm. Therefore, 2778 of these pairs of disks can be piled to 1 m. The diameter of the disc is 120 mm, so one side of length of the desk holder becomes 140mm.

Then, 49 disc holders can be placed in 1 m^2. As a result, 136,122 pair of discs will be packed into 1 m^3. As mentioned above, one set can generate 0.236 W, therefore, the total electric power that a 1-m^3 box of the new electrostatic generator can generate is 32.125 kW.

An ordinary home needs an electric power of 3 kW. Therefore, a 45.5-cm^3 box of this new electrostatic generator can supply the needed electric power.

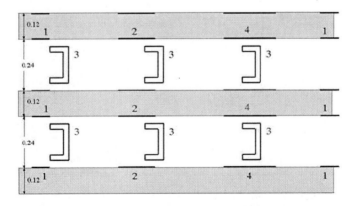

Figure 41. Schematic drawing of the piled discs of the new electrostatic generator.

Nevertheless, an electric car will require much more power. The charge density of the electret used in the simulation is 0.1 mC/m^2. The maximum charge density of the electret that we can use today is 1.0 mC/m^2 [8]. As a result, the generated electric power is 3212.5 kW/m^3 (explained in the following paragraph). This amount of electric power can easily run a car.

The electric field intensity is directly proportional to the charge density of the electrets. Additionally, the quantity of the induction charge of the charge carrier is directly proportional to the electric field intensity. Therefore, if the charge density increases by a factor of 10, the electric field intensity and the induction charge also increase by the same factor. As a result, the electrostatic force increases by a factor of 100. Finally, the generated electric power increases by a factor of 100 as well.

Electrets cannot produce a charge density greater than 1.0 mC/m^2. However, ferroelectric substances can usually produce charge densities of 10 mC/m^2. This material can generate an electric power of 321250 kW/m^3. The maximum charge density of ferroelectric material that we can use today is 160 mC/m^2 [9]. As a result, the generated electric power is 82240000 kW/m^3.

In the above-described calculation, it was assumed that the air resistance that arises with the high rotational speed of the charge carrier disc is zero. This assumption can be made valid by making the inside of the set a vacuum. A vacuum has three other advantages as well. One of them is that the corona discharge never occurs. The second is that the charge on the charge carrier will never leak into the air moisture. If either of these occurs, the electret charge will be quickly neutralized. The third advantage is that oxidation of the surface of the charge carrier never occurs. If this occurs, the electrostatic force does not act on the surface of the charge carrier perpendicularly. As a result, the ACF decreases.

In the above-mentioned calculation, it was assumed that the friction resistance that arises due to the high rotational velocity of the charge carrier disc is zero. This can be a valid assumption if the disc floats in the air by electrostatic force. The corresponding details are clarified in my Japanese patent, which will be open soon [10].

4.6. CONCLUSION

(1) The differences in the principles of the traditional electrostatic generator and the new electrostatic generator employing the ACF have been clarified. The former needs a mechanical force to lift up a

charge carrier, but the latter can lift up a charge carrier by using only an electrostatic force, without the need for a mechanical force.

(2) The electric energy that will be generated by one cycle by a minimum unit of the new generator is expected to be 1.37E-7 J.

(3) A concrete design and a manufacturing method of the new generator have been proposed. It can be made by conventional etching methods and corona discharge.

(4) The performance of the new generator is predicted to be 32 kW/m^3 when the charge density of the electrets is 0.1 mC/m^2.

4.7. OTHER APPLICATIONS OF THE ACF

Application of the ACF includes not only an electrostatic generator, but also an electrostatic transporter, an electrostatic accelerator, and an electrostatic motor. These three applications can be realized by an apparatus that has a structure similar to that of the above-mentioned electrostatic generator.

Chapter 5

A Miracle Charge Carrier
that Can Move Forward in
a Reverse Field

5.1. Background

As mentioned above, the basic theory of the new electrostatic generator is explained with Figure 33. The charge carrier is forced to the back by the electrostatic force in the reverse field. In Figure 33, Fe2 represents this force. If the direction of this force is reversed and the charge carrier is forced to move forward in the reverse field, the new electrostatic generator can generate a greater electric power.

A negatively charged charge carrier usually experiences a backward electrostatic force in a reverse electric field. Therefore, if the carrier moves forward in a reverse field, it is a miracle. It is an unbelievable phenomenon.

However, we know that a yacht can move forward as a result of wind power in a head wind. This phenomenon can be explained by the shape and the direction of the sails of the yacht. Therefore, if we select a suitable shape for the charge carrier, this miracle may occur.

I have attempted to solve this phenomenon, and I found a solution by simulation. I cannot yet confirm this simulation result with a real experiment. The shape of the charge carrier is complex; therefore, I can not make it correctly with handmade methods.

Thus, I will present only the simulation result now.

5.2. SIMULATION RESULT

Figure 42 shows a conception diagram of the miracle charge carrier.

In Figure 42, mark 3 shows the charge carrier. Marks 4 and 2 show the upper electrode and the lower electrode, respectively. When a large positive voltage is applied to the upper electrode and the lower electrode is grounded, then a forward electric field for a negative charge is formed between the electrodes. When the upper electrode is grounded and a large positive voltage is applied to the lower electrode, then a reverse electric field for a negative charge is formed between the electrodes. Figure 42 shows this state.

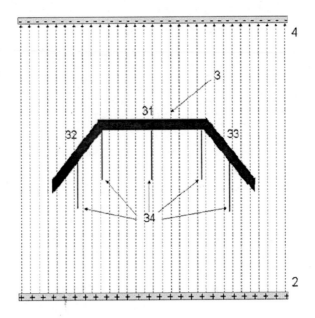

Figure 42. A conception diagram of the miracle charge carrier.

The miracle charge carrier 3 consists of a main plane 31 and two side planes 32 and 33 and five shield members 34, as shown in Figure 42. The direction of the main plane is perpendicular to the electric field, which is drawn as dotted lines in Figure 42. The direction of the two side planes is diagonal to the electric field. The direction of the shield members is parallel to the electric field. The shield members are attached to the back of the main and side planes.

They can shield the back of these three planes from the electric field. As a result, the electrostatic forces that act on the back of these three planes cancel each other.

(Caution: In reality, the electric line of force bends at the circumference of the charge carrier. However, it is drawn as a straight line for simplification in Figure 42.)

Figure 43 shows the detailed measurements of this charge carrier. The electrostatic force that acts on this charge carrier was simulated with these measurements.

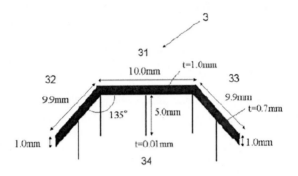

Figure 43. Detailed measurements of the miracle charge carrier.

This charge carrier has ten surfaces and three edges. Thus, thirteen electrostatic forces act on the ten surfaces and the three edges. This situation is drawn schematically in Figure 44.

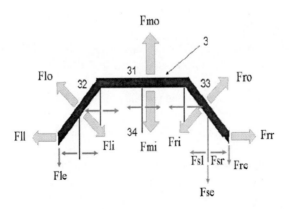

Figure 44. Schematic drawing of the thirteen forces that act on the ten surfaces and the three edges of the miracle charge carrier.

The thirteen electrostatic forces are the following forces.

> Fmo acts on the outside of the main plane.
> Fmi acts on the inside of the main plane.
> Flo acts on the outside of the left side plane.
> Fli acts on the inside of the left side plane.
> Fro acts on the outside of the right side plane.
> Fri acts on the inside of the right side plane.
> Fll acts on the left vertical surface of the left side plane.
> Frr acts on the right vertical surface of the right side plane.
> Fsl acts on the left surfaces of the five shield members.
> Fsr acts on the right surfaces of the five shield members.
> Fle acts on the edge of the left side plane.
> Fre acts on the edge of the right side plane.
> Fse acts on the edges of the five shield members.

These thirteen electrostatic forces were simulated by the following simulation method.

The simulation was done with the XY bi-dimensional finite difference method. Figure 45 shows the cell layout (mesh) used for the simulation. In this simulation, the charge carrier was given a charge of -3.0 nC, and it was placed in an electric field of ±2.08 MV/m.

In Figure 45, the measurements show the width and height of each cell, and the numbers shows the number of cells. The individual lines that outline the cells are eliminated for simplification.

Figure 46 shows the simulation result.

The following results are apparent in Figure 46.

(1) The inside force of the main plane (Fmi) and those of the two side planes (Fli, Fri) decrease largely with the shield members. The outside force of the main plane (Fmo) decreases by half with the shield members.

(2) The outside forces of the two side planes (Flo, Fro) increase slightly with the shield members.

(3) The left and right forces of the shield members (Fsl, Fsr) appear strongly with the shield members.

(4) The edge forces of the three edges (Fle, Fre, Fse) are small.

(5) The left force of the left side plane (Fll) and the right force of the right side plane (Frr) decrease slightly with the shield members.

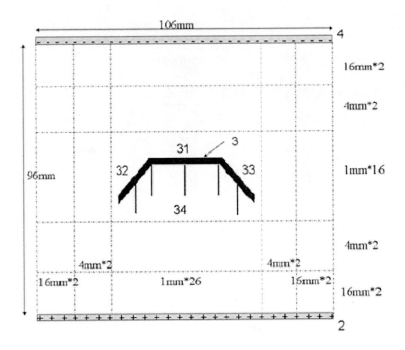

Figure 45. A design of cells (mesh) for simulating the electrostatic force that acts on a miracle charge carrier in a parallel electric field.

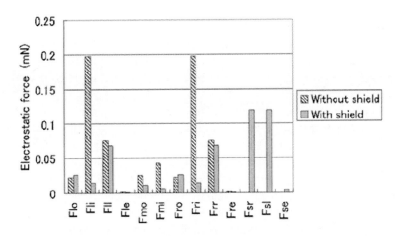

Figure 46. Simulation results of the thirteen electrostatic forces that act on the thirteen regions of the miracle charge carrier.

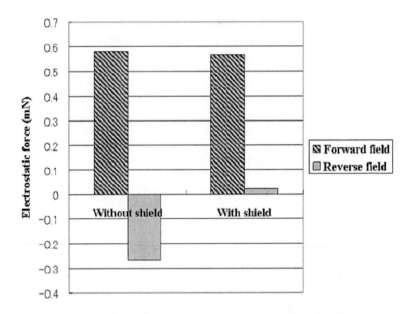

Figure 47. The electrostatic force that acts on the miracle charge carrier in a forward or reverse electric field with or without shield members.

The left and right forces of the shield members (Fsl, Fsr) cancel each other, because they act in opposite direction. The left force of the left side plane (Fll) and the right force of the right side plane (Frr) cancel each other for the same reason. The parallel component of the outside force of the left side plane (Flo) and the parallel component of the outside force of the right side plane (Fro) also cancel each other for the same reason. Finally, the parallel component of the inside force of the left side plane (Fli) and the parallel component of the inside force of the right side plane (Fri) cancel each other for the same reason.

I then summed all of the outside forces of the main plane (Fmo) and the vertical component of the outside forces of the two side planes (Flo, Fro); this sum is the total upper force. Additionally, I summed up the inside force of the main plane (Fmi) and the vertical component of the inside forces of the two side planes (Fli, Fri) and the three edge forces(Fle, Fre, Fse). This is the total lower force. Finally, I summed up the two total forces. This is the final force that acts on the miracle charge carrier. Figure 47 shows the four final forces that act on the miracle charge carrier in a forward or reverse electric field with or without shield members.

The following results are apparent in Figure 47.

(1) The electrostatic force that acts on the miracle charge carrier in the forward electric field decreased slightly when five shield members were attached to the back side of the carrier.

(2) In contrast, the force in the reverse electric field decreased significantly, and the direction was reversed.

This result means that a miracle charge carrier that has a negative charge can move forward to a negative potential in a reverse electric field.

5.3. CONCLUSION

It was found by a simulation that, if the back side is electrically shielded, a charged specially shaped conductor can progress from a low potential to a higher potential for the charge by an electrostatic force that is generated by a electric field.

CONCLUSION

1) The electrostatic force that acts on asymmetric conductors was simulated, and the following three asymmetric forces were found.

    A.   The asymmetric force (AF), which acts on a neutral asymmetric conductor in a parallel electric field. Mathematically, this force must be equivalent to the gradient force that acts on a neutral symmetric conductor in a convergent electric field.

    B.   The Asymmetric Image Force (AIF), which acts on a charged asymmetric conductor near an electrode. The magnitude of the force changes with the shape and direction of the conductor.

    C.   The Asymmetric Coulomb Force (ACF), which acts on a charged asymmetric conductor in a parallel electric field. When the direction of the field is reversed, the absolute value of the force changes, even if the absolute value of the field remains the same.

2) The existence of AF and ACF was confirmed experimentally.

3) The causes these three forces were made clear. The asymmetric conductors that experience these three forces have perpendicular planes and parallel planes, with respect to the electric field, and a large area of the perpendicular planes of one side are electrically shielded by the parallel planes. Only a small amount of charge can gather at the small area of the non-shielded perpendicular plane on one side, but many charges gather on the large parallel plane and do not contribute to the effective force.

4) A concrete design and a production method for a new electrostatic generator driven by ACF were devised, and the generator's performance was predicted. The performance can be used to simultaneously solve the environmental problem and the energy crisis.

5) A miracle charge carrier was devised. This specially shaped carrier can move from a low potential to a higher potential for the carried charge by the electrostatic force that is generated by an electric field.

APPENDIX 1. AN EXPLANATION OF THE SIMULATION METHOD (A BI-DIMENSIONAL AXI-SYMMETRIC FINITE DIFFERENCE METHOD)

Here, I will explain how to simulate the electrostatic force that acts on a non-charged cylinder that is floating in an electric field (see Figure 2). This simulation is performed by using a bi-dimensional axi-symmetric finite difference method.

First, the simulation target space between the two circle electrodes was divided into small parts, as schematically shown in Figure 48.

The space was divided into 5 parts in the R direction and 12 parts on the Z axis in this simulation, although, in Figure 48, it is only divided into 5 on the Z axis for clarity. I call this divided small space a part. As shown in Figure 48, the shape of the 12 parts on the Z axis is cylindrical and the shape of the other 48 parts is a pipe shape. The bold line circles show the edges of the conductor parts, and the normal line circles are the edges of the insulator (air) parts. The potential along these circles is the same everywhere. Each part has three (cylinder) or four (pipe) surfaces, that is, the left surface, the right surface, the outer circumference surface, and the inner circumference surface.

The surface charge density and the electric field intensity of each surface are determined by a simulation, and the electrostatic force acting on the surface is calculated by those two values.

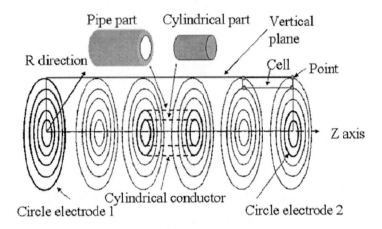

Figure 48. Schematic drawing of the space between the electrodes for the axi-symmetric finite difference method.

Point number

1	8	15	22	29	36	43	50	57	64	71	78	85	92
2	9	16	23	30	37	44	51	58	65	72	79	86	93
3	10	17	24	31	38	45	52	59	66	73	80	87	94
4	11	18	25	32	39	46	53	60	67	74	81	88	95
5	12	19	26	33	40	47	54	61	68	75	82	89	96
6	13	20	27	34	41	48	55	62	69	76	83	90	97
7	14	21	28	35	42	49	56	63	70	77	84	91	98

Figure 49. A design of cells (mesh) for simulating the electrostatic force that acts on a non-charged cylinder in a convergent electric field.

This is a very simple simulation. Therefore, we can describe the simulation air space in three dimensions. However, in the case of a complicated simulation, it may be difficult to describe it in three dimensions. Thus, we usually choose one plane that includes the Z axis of the target space. Because the space is axi-symmetric, we describe only one side of the

Z axis. For example, only the upper part of the vertical plane was converted into a bi-dimensional picture, which is shown in Figure 49. In this conversion, a section of the parts (the pipe and the cylinder) was changed into a cell. Furthermore, the circumference line of the parts (the pipe and the cylinder) was changed to a point. Figure 49 shows the details of the converted vertical half plane.

The mesh is usually cut over lager air space that include the simulation target, as shown in Figure 49. In the figure, a normal digit represents a lattice point number. The potential of each lattice point is calculated by the simulation. For example, the potential of lattice point 52 is calculated with the following formula from the electric potential of the four neighboring lattice points.

$$V52 = \frac{V51 + V53 + V45 + V59}{4} \tag{10}$$

The same formula is used at all lattice points. For this coalition linear equation, the potential is solved using the Gaussian method of elimination from a lattice point that is already known. For example, a value of +600 V is given to lattice points 9, 10, 11, 12, 13, and 14 because electrode 1 includes lattice points 9, 10, 11, 12, 13, and 14. A value of 0 V is given to lattice points 96, 97, and 98 because electrode 2 includes these lattice points. Because the cylinder includes 21 lattice points from 33 to 77, the potential of these points are forced to be the same.

In Figure 49, the oblique digits S1 - S10 represent the ten surfaces of the cylinder. The electric field intensity, En, of the surface area of the cylinder is calculated using the potentials of the four corner points of the cell. For instance, the electric field intensity E1 of surface 1 is calculated from the potential of the four points V27, V28, V34, and V35, and the width of cells, w, is determined by formula (11):

$$E1 = \frac{(V27 + V28)/2 - (V34 + V35)/2}{w} \tag{11}$$

The surface charge density of each surface δ_n is calculated from the field intensity E_n of the surface and the vacuum permittivity ε_0, using formula (12):

$$\delta_n = \varepsilon_0 E_n, \tag{12}$$

The charge on each surface q_n is calculated from the surface charge density δ_n and area of the surface S_n, using formula (13):

$$q_n = \delta_n S_n, \tag{13}$$

The electrostatic force F_n acting on each surface is calculated using formula (14):

$$F_n = \frac{q_n E_n}{2} = \frac{\varepsilon_0 S_n E_n^2}{2} \tag{14}$$

The above is a simple explanation of the simulation method. However, in practice, various problems can arise. The details of the simulation method are described in references [11], [12], [13].

Figure 49 shows the design of cells (mesh) used for this simulation. F1 is the sum of the electrostatic forces that act on the left side surfaces S1 and S2 of this cylinder. F2 is the sum of the electrostatic forces that act on the right side surfaces S9 and S10 of this cylinder. Finally, F3 is the sum of the electrostatic force that act on the circumference surfaces S3, S4, S5, S6, S7, and S8 of this cylinder. The electrostatic force Fe that acts on the cylinder was calculated using formula (2):

$$Fe = F1 + F2 \tag{2}$$

The force F3 that acts on the circumference surface of the cylinder was not included in formula (2) because it cancels out at an interval of 180 degrees and ultimately becomes zero.

APPENDIX 2. RELATIONSHIP BETWEEN THE APPROXIMATE FORMULA FOR THE GRADIENT FORCE AND THE SIMULATION METHOD OF THE ELECTROSTATIC FORCE ACTING ON A NON-CHARGED CYLINDER IN A CONVERGENT FIELD

This axi-symmetric bi-dimensional finite difference method cannot deal with a sphere; therefore, a spherical conductor with radius r is replaced by a cylinder that envelops the spherical conductor. The radius of both side surfaces of the cylinder is r, and the width is 2r (see Figure 50).

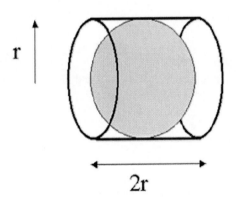

Figure 50. A spherical conductor (radius = r) and a cylinder (radius = r, width = 2r).

The electrostatic force F_1 that acts on the left surface of the cylinder is calculated by formula (15):

$$F_1 = \frac{\varepsilon_0 S_1 E_1^{\,2}}{2},$$ (15)

where ε_0 is the vacuum permittivity, S_1 is the area of the left surface and E_1 is the field intensity on the left surface.

The electrostatic force F_2 that acts on the right surface is calculated with formula (16):

$$F_2 = \frac{\varepsilon_0 S_2 E_2^{\,2}}{2},$$ (16)

where ε_0 is the vacuum permittivity, S_2 is the area of the right surface and E_2 is the field intensity on the right surface. The magnitude of E_2 is larger than that of E1, because this field converges to the right.

The total electrostatic force Fe is calculated with formula (2):

$$F_e = F_1 + F_2.$$ (2)

The direction of F_1 is to the left (-) while that of F_2 is to the right (+), and formula (2) is transformed into formula (17),

$$F_e = -\frac{\varepsilon_0 S_1 E_1^{\,2}}{2} + \frac{\varepsilon_0 S_2 E_2^{\,2}}{2}$$ (17)

The areas of the left and right surfaces of the cylinder are the same and equal to πr^2, so formula (17) becomes formula (18):

$$F_e = \frac{\varepsilon_0 \pi r^2 (E_2^{\,2} - E_1^{\,2})}{2}.$$ (18)

The width of the cylinder w is multiplied by the denominator and the numerator in formula (18):

$$F_e = \frac{\varepsilon_0 \pi r^2 w (E_2{}^2 - E_1{}^2)}{2w} \ .$$

(19)

The width of the cylinder w is equal to 2r, and formula (19) becomes formula (20):

$$F_e = \frac{\varepsilon_0 \pi r^3 (E_2{}^2 - E_1{}^2)}{w}$$

(20)

$(E_2{}^2 - E_1{}^2)/w$ represents the gradient of the field intensity squared at the right and left surfaces of the cylinder, so formula (20) becomes formula (21):

$$F_e = \varepsilon_0 \pi r^3 \nabla E^2$$

(21)

This formula is derived based on the assumption that the electric field intensity is not changed when the cylinder is placed in the electric field, but it actually becomes stronger because the electric field concentrates in the left and right surfaces of the cylinder. The concentration rate is not known, but, if it is $\sqrt{2}$, then formula (21) is transformed into formula (22):

$$F_e = 2\varepsilon_0 \pi r^3 \nabla E^2.$$

(22)

Formula (22) is the same as the gradient force approximate formula (1).

Therefore, the simulated electrostatic force acting on the cylinder in the convergent field must be the gradient force.

Nevertheless, the shape of the conductor is not a sphere, it is a cylinder. Therefore, I now attempt to simulate an electrostatic force acting on an analogous spherical conductor in a convergent field.

Figure 51 shows a spherical conductor (dotted line) and a piled six-disk conductor as an approximately spherical conductor.

The radius of the sphere is R, and the radius of the first left and first right disks of the piled conductor is R/2. The width of the piled conductor is $\sqrt{3}$ R. E_{10} and E_{20} are the original electric field intensity, and E_1 and E_2 are the changed electric field intensity after the piled conductor is placed in the convergent field. The field concentration rate on the first left and first right surfaces of the piled conductor is assumed to be 3.0. It is actually about 2.7

in the simulation of the piled three-disk conductor that was described in Chapters 1-3, but considering the electrostatic force acting on the second and third surfaces of the piled conductor (see Figure 7), it is assumed to be 3.0 here.

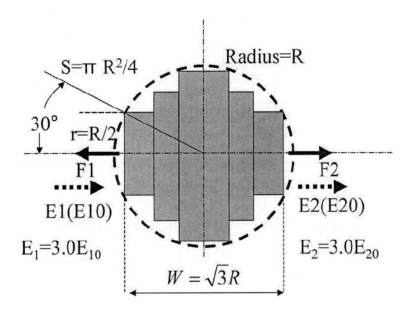

Figure 51. A spherical conductor (radius = R) and a piled six-disk conductor as an approximately spherical conductor (radius of the first left and right surfaces = R/2, width = $\sqrt{3}$ R).

Caution: As mentioned above, I assumed that the concentration rate of the electric field on the surface of the cylinder that is shown in Figure 50 is $\sqrt{2}$, but I use 3.0 here, because the radius of this cylinder is only half of that of the cylinder that is shown in Figure 50).

The electrostatic forces that act on the second left and right surfaces and on the third left and right surfaces are neglected (see Figure 7).

Therefore, the piled six-disk conductor can actually be replaced with a cylinder whose radius is R/2 and whose width is $\sqrt{3}$ R.

The electrostatic force Fe that acts on the cylinder is calculated with formula   (20)   with   r=R/2,   w= $\sqrt{3}$ R, $E_1 = 3.0 \times E_{10}$,   and

$E_2 = 3.0 \times E_{20}$ are substituted into formula (20), which then becomes formula (23):

$$F_e = \frac{\varepsilon_0 \pi (\frac{R}{2})^2 \sqrt{3} R((3.0 \times E_{20})^2 - (3.0 \times E_{10})^2)}{2w} \tag{23}$$

$$= \frac{9\sqrt{3}\varepsilon_0 \pi R^3 (E_{20}^{\ 2} - E_{10}^{\ 2})}{8w}$$

$$= \frac{1.95\varepsilon_0 \pi R^3 (E_{20}^{\ 2} - E_{10}^{\ 2})}{w}.$$

$(E_{20}^2 - E_{10}^2)/w$ represents the gradient of the field intensity squared at the left and right surfaces of the cylinder, and formula (23) is transformed into formula (24):

$$F_e = 1.95\varepsilon_0 \pi R^3 \nabla E^2 \tag{24}$$

$$\doteqdot 2\varepsilon_0 \pi R^3 \nabla E^2$$

Formula (24) is the same as the gradient force approximate formula (1). Of course, this is not an exact simulation of a spherical conductor. An exact simulation would require a three-dimensional simulation method. Therefore, an exact simulation of a spherical conductor is required in further work.

REFERENCES

[1]   Handbook of electrostatic (Japan) (1998) p.1196.

[2]   K. Sakai, An overlooked electrostatic force that acts on a non-charged asymmetric conductor in a symmetric (parallel) electric field, *J. Electrostat.* 67 (2009) 67-72.

[3]   K. Sakai, An experimental result which confirm the fourth electrostatic force, Proceedings of ESA Annual Meeting on Electrostatics 2008 (2008).

[4]   The fourth edition of physics and chemistry dictionary (Japanese), p.309.

[5]   L. B. Schein, Electrostatic Proximity force, Toner Adhesion, and a New Electrophotographic Development System, Proc. ESA/IEJ/IEEE-IAS/SFE Joint Conference on Electrostatics 2006 68-75.

[6]   K. Sakai, The electrostatic force that acts on the charged asymmetric conductor in normal and reverse electric fields, Proceedings of 2009 Electrostatics Joint Conference (2009).

[7]   Handbook of electrostatic (Japan) (1998) p.962.

[8]   T. Tsutsumino, Y. Suzuki, N. Kasagi, Y. Sakane, Seismic power generator using high-performance polymer electret, Proc. Int. Conf. MEMS'06, Jun. 2006, Istanbul, pp.98-101.

[9]   Kageyama, K., Kato, H. and Watanabe, S.: Fabrication and Evaluation of Electret Condenser Microphone using Poled Ferroelectric Ceramics, Journal of Solid Mechanics and Materials Engineering, *Journal of Solid Mechanics and Materials Engineering*, vol. 2, No. 7, pp 877-887 (2008).

[10] K. Sakai, Japanese patent during an application, *Number* 2008-289259.

[11] Yasuyuki Matsubara, Improved Finite Difference Expression for
       Laplace's and Poisson's Equations on the Surface of Dielectrics, *J.
       Electrostatics Japan* 16 (1992) 440-442.
[12] Yasuyuki Matsubara, A Guide to the Calculation of Electric Field Part
       □ Application of the Finite Difference Method to Electric field in an
       Oil Tank, *J. Electrostatics Japan* 16 (1992) 530-538.
[13] Yasuyuki Matsubara, A Method to Calculate the Potential of
       Ungrounded Conductors, Proceedings of the Institute of Electrostatics
       Japan (1994) 181-184.

INDEX

A

accelerator, 16, 54
airplane, viii
applications, 23, 54
authors, vii

C

carrier, vii, viii, 25, 41, 42, 43, 44, 45,
  46, 47, 48, 50, 51, 52, 53, 54, 55, 56,
  57, 58, 59, 60, 61, 64
cell, 11, 19, 26, 34, 45, 58, 67
charge density, 44, 53, 54, 65, 67, 68
charged powder, vii
clarity, 65
CO_2, 43
concentrates, 71
concentration, 71, 72
conception, 56
concrete, 41, 47, 48, 54, 64
conductor, viii, 1, 2, 3, 4, 5, 6, 7, 8, 9, 11,
  13, 14, 16, 17, 19, 20, 22, 23, 26, 27,
  29, 32, 34, 35, 37, 38, 45, 61, 63, 65,
  69, 71, 72, 73, 75
conductors, 10, 16, 34, 35, 36, 38, 63
consensus, 36
conservation, viii
convergence, 7

conversion, 67
copper, 13, 47, 50
corona discharge, 42, 50, 53, 54

D

damages, iv
density, 44, 53, 54, 65, 67, 68
discs, 47, 48, 51, 52
distribution, 21, 22, 36, 37, 38, 39
drawing, 52, 57, 66

E

electric charge, vii
electric field, vii, viii, 1, 3, 4, 5, 6, 8, 9,
  12, 13, 15, 16, 17, 21, 25, 26, 27, 28,
  29, 30, 31, 32, 33, 34, 36, 37, 38, 42,
  44, 46, 50, 53, 55, 56, 57, 58, 59, 60,
  61, 63, 64, 65, 66, 67, 71, 72, 75
electrodes, 3, 4, 11, 13, 14, 16, 17, 27,
  29, 30, 34, 44, 45, 46, 47, 48, 50, 51,
  52, 56, 65, 66
electron, 44
electrons, 36, 37, 42
energy, viii, 42, 43, 44, 45, 46, 47, 51,
  54, 64
equipment, 28
etching, 47, 50, 54

F

floating, 15, 32, 65
foils, 50
formula, vii, viii, 2, 3, 4, 8, 15, 19, 20, 25, 34, 45, 46, 47, 51, 67, 68, 70, 71, 72, 73
freezing, 50
friction, 53

G

graph, 34
gravity, 15, 21, 32
grounding, 46

H

height, 15, 20, 32, 44, 58
hypothesis, 1, 4, 7, 17

I

ideal, 2, 47
image, 1, 19, 20, 21, 23, 50
induction, 26, 29, 37, 44, 45, 46, 47, 48, 49, 50, 53
infinite, 43
injury, iv
interference, 49
interval, 4, 9, 22, 37, 45, 68
ions, 42, 50

J

Japan, viii, 75, 76

L

lifetime, 43
limitation, vii

line, 3, 14, 42, 48, 50, 52, 57, 65, 67, 71

M

manufacturing, 41, 50, 54
moisture, 53
movement, viii

O

oil, 47, 50
oils, 50
order, 3
oxidation, 53

P

parallel, 1, 2, 4, 5, 6, 7, 8, 9, 10, 12, 13, 14, 15, 16, 17, 26, 37, 38, 56, 59, 60, 63, 75
performance, 41, 50, 54, 64, 75
permission, iv
permittivity, 19, 67, 70
photographs, 30
physics, 75
pitch, 51, 52
polycarbonate, 13, 50
polymer, 75
power, 13, 14, 16, 41, 51, 52, 53, 55, 75
prediction, 41
production, 64
program, 2, 4, 11

R

radius, 4, 20, 26, 48, 69, 71, 72
reality, 57
reason, 60
recommendations, iv
recovery, 42, 44, 45, 46, 47, 49, 50, 51, 52
region, 37

resistance, viii, 53
respect, 10, 63
rights, iv

S

shape, vii, viii, 1, 4, 7, 10, 11, 17, 19, 25,
    43, 55, 63, 65, 71
silk, 13
simulation, vii, viii, 2, 3, 4, 7, 8, 11, 13,
    14, 16, 17, 19, 20, 21, 23, 25, 26, 27,
    30, 34, 36, 38, 45, 48, 53, 55, 58, 61,
    65, 66, 67, 68, 72, 73
space, vii, 47, 65, 66, 67
specific gravity, 15
specific surface, 8
speed, 48, 51, 53
strength, 15
supply, 13, 14, 41, 52
surface area, 5, 10, 11, 67

T

tensile strength, 15
transportation, 16

U

uniform, vii

V

vacuum, 53, 67, 70
variables, 7
velocity, 51, 53

W

wind, viii, 55